WHAT IS HERPETOLOGY?

Herpetology is the scientific study of **amphibians** and **reptiles**, including frogs.

The scientists who study frogs are called **HERPETOLOGISTS.**

Words that are tricky to understand are in **bold**. Find out what they mean in the glossary.

Words that are difficult to say are in *italics.* Find out how to say them at the back of the book.

WHAT CAN FROGS TELL US ABOUT OUR PLANET?

DISCOVER THE SCIENCE BEHIND **HERPETOLOGY**
(huh-puh-TOH-luh-jee)

Written by Eliza Jeffery
Illustrated by Valeria Abatzoglu

Frogs tell us so much about our planet. By keeping an eye on their numbers and how they behave, we can see how our world is changing. Frogs have a big part to play in keeping nature healthy. So what can frogs tell us about our planet?

Frogs can give early warning signs that a **habitat** is in danger because of their skin! Their skin is moist and thin, allowing water and **gases** to pass through it easily. It's how frogs drink and breathe, and the reason they react so quickly to the world around them.

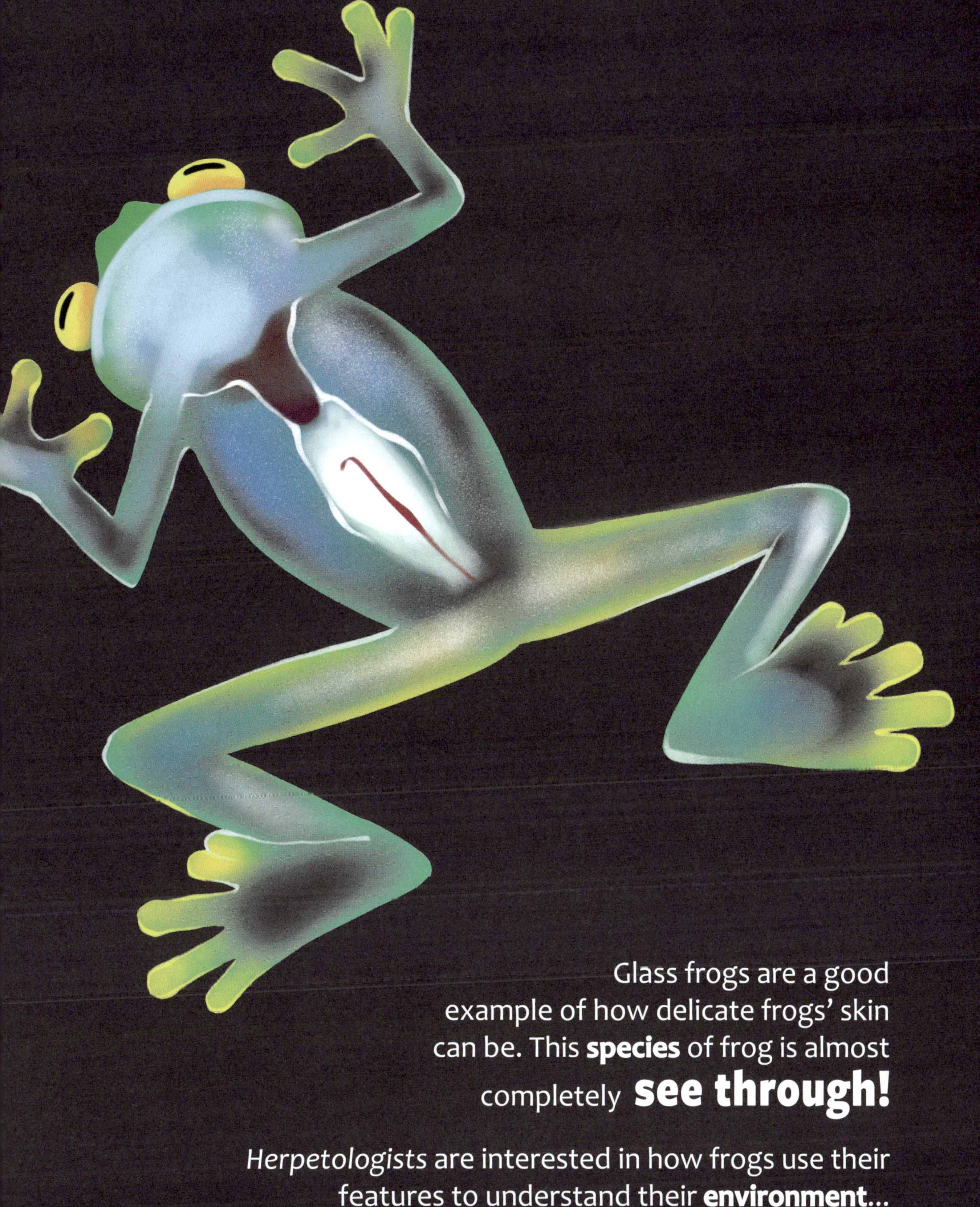

Glass frogs are a good example of how delicate frogs' skin can be. This **species** of frog is almost completely **see through!**

Herpetologists are interested in how frogs use their features to understand their **environment...**

Pollution often ends up in natural areas, and because of their sensitive skin, it affects frogs first. This means frogs are often the first animals to show signs that all is not well.

Pollution that finds its way into watery habitats can affect all stages of a frog's life. It can damage their eggs, leading to fewer hatching. Those that do hatch **may grow extra legs!**

Frogs are seen as **indicator species** of **climate change.** Herpetologists have noticed that over time, frogs have had to **adapt** to more extreme weather.

For example, wood frogs have adapted to live at the top of snowy mountains. To help them stay alive, wood frogs can **freeze their own bodies!**

In warmer climates, the giant monkey tree frog has adapted to high temperatures by covering itself in a sun-protecting liquid made from its body. Just like humans, this frog **puts on sunscreen!**

As temperatures rise, more frogs are affected. Their thin skin makes them more sensitive to heat than other animals. By watching frogs and their habits, scientists can see climate change in action.

Not only do frog populations tell us a lot about our planet, they can tell us about ourselves too! Herpetologists have discovered that African clawed frogs are similar to humans in how their bodies react to some illnesses. By finding these connections between frogs and humans, scientists can better understand human illnesses.

Not only can frogs teach us about human illnesses, but some can help cure them too!

Toads are a type of frog. The Asiatic toad and Asian common toad produce liquids which are used to treat allergies, soothe pain, and cure some illnesses. Toads and frogs may look small, but they're a big deal. These animals are

saving lives!

Frogs have a really important place in the **food chain** because they can be both **predator** and **prey.** Without frogs, some species would go hungry, and others would **overpopulate.**

If the number of animals in an area decrease, scientists check on frog populations to learn more!

Frogs also protect the natural spaces they live in. The Wallace's flying frog eats lots of the insects that ruin plants that act as homes for other animals. By protecting plants, these little creatures protect the wider **ecosystem** we all live in.

If frogs are struggling to survive, it's a warning sign to herpetologists that the whole environment may be in danger.

Unfortunately, frogs are in danger because of humans. Newly built roads often get in the way of toad's **migration** routes, which they follow every year. This puts them in danger every time they try to cross!

FRO 66Y

Herpetologists know how important frogs are, so they spend a lot of time learning how to protect their populations.

Not only does this help us understand our planet better, it helps us understand frogs better too. By keeping count of how many live in one area, herpetologists can figure out when frogs need some human help!

Thanks to herpetologists, we can better understand what frogs are telling us about the world we live in. There are so many different frogs all around the world, helping us to look after our planet…

Houston toad

Panamanian golden frog

Ornate horned frog

Fire-bellied toad

Green and black poison dart frog

Red-eyed tree frog

Blue poison dart frog

Tomato frog

Darwin's frog

White tree frog

Argentine horned frog

Impressive

FROGS

Now we know all about how frogs help our planet, what frogs are there that stand out from the crowd? Here's just a few of the most impressive ones!

Ancient GIANT!

Beelzebufo was the biggest frog that ever lived! It was as big as a beach ball, and lived millions of years ago. It is nicknamed the "Devil Toad"!

Unusual EGGS!

The *Surinam* toad is an unusual type of frog – the females carry their eggs inside pockets in their backs! When the eggs are ready to hatch, the young frogs burst out from the skin in search of food.

Poisonous and DEADLY!

Although the golden poison frog is very small, it has enough **poison** to harm lots of creatures!

Super SMALL!

The *New Guinea Amau* frog is so tiny it can sit on a coin! It's the smallest frog in the world, and can only be found in New Guinea.

Flying EXPERTS!

Wallace's flying frog doesn't really fly, but can glide from tree to tree using its big, webbed feet! It likes to live high up in rainforest trees.

Fantastic

FROG FACTS

Frogs can tell us lots of things about the world around us, but what else is there to know about these unique, little creatures?

ALL TOADS ARE FROGS, BUT NOT ALL FROGS ARE TOADS!

Toads tend to have drier, bumpier skin, and spend more time on land, whereas frogs much prefer to be by the water's edge.

FROGS WERE THE FIRST LAND ANIMALS WITH VOCAL CORDS!

They use their voices to communicate with other frogs, or when they feel they are in danger, producing a variety of croaks and calls.

FROGS ARE EXCELLENT JUMPERS!

Launched by their long legs, many frogs can leap more than 20 times their body length! This makes them excellent hunters.

FROGS USE THEIR EYEBALLS TO SWALLOW PREY!

When a frog swallows its prey, it blinks, which pushes its eyeballs down on top of the mouth. This helps it to push the food down its throat!

FROGS EAT THE SKIN THEY SHED!

Because a frog's skin is full of goodness, a frog will often eat the skin it has just shed, **recycling** the valuable **nutrients**.

GLOSSARY

Adapted – when a living thing has developed special features or skills to help it survive in its environment (see below).

Amphibians – cold-blooded animals that live both in water and on land during different stages of their life cycle. Examples include frogs, toads, and salamanders.

Climate change – a change in the weather conditions over a long time.

Ecosystem – all the living and non-living things that exist together within an area.

Environment – everything that is around us.

Food chain – the order in which different animals eat each other to survive.

Gases – tiny, usually invisible, particles in the air.

Habitat – the place where animals and plants live.

Indicator species – species that can provide information about the health of an ecosystem (see left).

Migration – the movement of animals from one area to another.

Nutrients – substances or ingredients that plants and animals need to live and grow.

Overpopulate – when a species' (see right) population grows too large for the environment (see left) to support it.

Poison – a substance that can kill or hurt a person or animal.

Pollution – adding harmful materials into the environment (see left).

Predator – an animal that hunts other animals for food.

Prey – an animal that is hunted by other animals for food.

Recycling – taking old things and making them into new things.

Reptiles – a group of cold-blooded animals, including snakes, lizards, and dinosaurs.

Species – a group of living things that share characteristics and features, and can have babies together. For example, goliath frogs and poison dart frogs are different species.

HOW DO I SAY?

Beelzebufo
bee-zeh-BUH-foe

Herpetologists
huh-puh-TOH-luh-jists

Herpetology
huh-puh-TOH-luh-jee

New Guinea Amau
nyoo GIN-ee ah-MOW

Surinam
SUOR-ruh-nam

THE BIG QUESTIONS ANSWERED

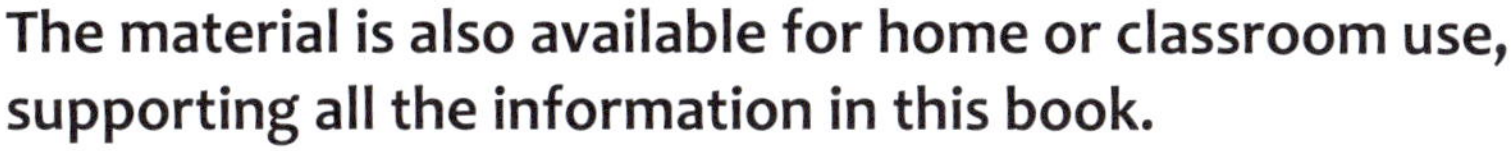

This is more than just a series of books; it is a complete resource. Accompanying each book is a variety of FREE material to engage curious kids with science.

www.thebigquestionsanswered.com

Use the QR code to visit the website, download free resources, and discover other books in the series.

On the website, find out incredible things about herpetologists, including what they do, some of their greatest discoveries, and the people who have made a difference in this field of science.

The material is also available for home or classroom use, supporting all the information in this book.

Teachers' & Parents' Resources
With discussion prompts and questions, extra information, and facts around key topics.

Young Herpetologists' Activity Pack
Fun activities for wannabe frog experts, including creative writing, drawing, word searches, and much, much more.

The Big Questions Answered is published by Beetle Books. Beetle Books is an imprint of Hungry Tomato Ltd.

First published in 2025 by Hungry Tomato Ltd
F15, Old Bakery Studios, Blewetts Wharf, Malpas Road, Truro, Cornwall, TR1 1QH, UK.

ISBN 9781835691427

A CIP catalog record for this book is available from the British Library.

With thanks to:
Editors: Millie Burdett and Holly Thornton
Senior Designer: Amy Harvey
The team at Beehive Illustration

Information in this book is up to date as of the time of writing.

Printed and bound in China.

Picture Credits:
(t = top, b = bottom, m = middle, l = left, r = right)
Shutterstock: Artsiom P 35mr; CGS Graphics 35tl; oleJohny 34bl; satoriphoto 33ml; Thorsteinn Asgeirsson 33br; Tunatura 32mr; RethaAretha 34mr; Wirestock Creators 35bl.